I0134839

FOOD TRUCK BUSINESS

The Practical Beginner's Guide To Building And Growing Your Business Profitably. Strategic Inclinations To Break Down Significant Mistakes And Make your Sales Last

David Fridman

© **Copyright 2022 David Fridman - All rights reserved.**

The content contained within this book may not be reproduced, duplicated or transmitted without direct written permission from the author or the publisher.

Under no circumstances will any blame or legal responsibility be held against the publisher, or author, for any damages, reparation, or monetary loss due to the information contained within this book. Either directly or indirectly.

Legal Notice:

This book is copyright protected. This book is only for personal use. You cannot amend, distribute, sell, use, quote or paraphrase any part, or the content within this book, without the consent of the author or publisher.

Disclaimer Notice:

Please note the information contained within this document is for educational and entertainment purposes only. All effort has been executed to present accurate, up to date, and reliable, complete information. No warranties of any kind are declared or implied. Readers acknowledge that the author is not engaging in the rendering of legal, financial, medical or professional advice. The content within this book has been derived from various sources. Please consult a licensed professional before attempting any techniques outlined in this book.

By reading this document, the reader agrees that under no circumstances is the author responsible for any losses, direct or indirect, which are incurred as a result of the use of information contained within this document, including, but not limited to, errors, omissions, or inaccuracies.

Table of Contents

David Fridman

Hi, I'm David.

Thank you for purchasing my book.

I have been running my food truck business for several years with great success and passion.

I started with a small food truck around the neighborhoods of New York City. It was tough and very challenging. I had to make huge sacrifices, financially and in terms of my relationship with my wife and loved ones.

However, all this was a great training ground for my path and my education.

In this book, you will find many insights into your own business and be given many ideas.

Cooking has always been my great passion, and I have taken 3 training courses.

Today I can earn more money than in my old salaried job, plus I am happy and peaceful with my family.

To do this kind of business, you need to be disciplined, practical, confident and above all, be very friendly. All our customers have a unique experience when they taste our products. So, we necessarily have to give them a perfect impression.

Feel free to contact me quietly at my personal email address,

dfridman.foodtruck@outlook.com

or, if you prefer, you can sign up for my private list by simply scanning the QR Code on this page.

I never send spam, and your email is safe; I will not share your email address.

Now let's get started. Happy reading.

Introduction

The food truck business is booming. If you want to get in early and get your foot in the door of the multi-billion-dollar industry, now is your chance. This business can be very lucrative if you are good at what you do. The best time to start looking into it is now before something else takes off!

The food truck business involves transporting different types of food from one location to another. Usually, these trucks work around a central point where they can spend their day cooking for people that order them or just want a snack. Other times they cook at the location. This fact makes it a viable business for anyone who loves to cook or people who just want to make a lot of money. Whatever you do, don't waste any time. Research the industry and figure out what you want to do, then go out there and do it!

So, what does it take to be successful in this type of business? First, you need to have some serious cooking skills. You will need to know how to whip up delicious food fast and efficiently for your customers. If your food does not taste good, people will stop returning and giving your business. So, you need to be able to show off your cooking skills. It would help if you also had some excellent business sense. It takes a lot of business sense to make this industry work for you. You will need to know what your customer wants, how much it will cost, and how much you want in return for this dish.

This is not a "make it big or fail" type of business, but you do have to put in some time and effort if you want it all. This is not typically a one-person operation, so think about getting some help or partners when starting this business.

There are so many benefits to owning a food truck that it's hard to list them all. They mostly go into the business that you hold and how flexible it is. For example, there are busier times of the day to be out and about. Things like happy hours and school nights work best for food truck owners. This is because more people like to eat at those times and be able to make it home before dinner or go out with their friends without having to worry about making it home on time.

Also, this business can be run from anywhere. Some days you may want to set up shop in one area, and other times you may want to try out new places. With a food truck, you can do this with ease. You are not confined to one location but can choose where you want to go. As far as restrictions go, there is none besides the cooking regulations that are set into place by local cities or states. You will have to do your research into this before diving into the food truck business because some laws need to be followed around certain areas of the country, so make sure you know what they are before doing any type of business there.

So how will you run your food truck? You can either have your truck work directly for you or let others use it. Either way, you will have to pay rent to the place where you are working. You also have to pay for insurance, maintenance, gas, food costs, and more. So, keep all of this in mind when deciding whether to get your own standalone business or just get someone else's food truck.

When starting a new business venture, you mustn't make any costly mistakes early to ensure success. The food truck business is no exception to this. Here are some things that you should consider when starting up your food truck.

- **Finding a niche:** The one thing that many people struggle with when coming up with a place for their business is not knowing who they are selling to. As stated above, you need to be aware of what your clientele wants for them to come back to you again. If you know who your customers want, then you can focus your efforts there and ensure that you do not fall short in any area.

- **Looking for a good location:** Location, location, location! This is the worst business in the world if you don't have a good location. You are not likely to make any money if you are far away from potential customers. You need to find a place with lots of people in it when you want to make money. How will you do this? One way is by using Google Maps. If you can't find a spot using this tool, then it may be time just to leave that particular area and try somewhere else!

- **Finding your workforce:** If you want to know how to start a food truck business, finding your workforce needs to be at the top of your list. You will need people to help you cook and sell everything you make. Working alone can be very demanding and stressful, so make sure that you find all the help you need within your niche. This is a great way to keep your customers coming back for more!

- **Getting the right equipment:** You need to ensure that you get the right equipment before getting your food truck up and running. You will need a refrigerator, stove, cabinets, tables/counters, grills, sinks, hot water tanks, and more. You will also have to have cupboards or other equipment where customers can pay for the food and store everything that they buy from you.

This book will detail each of these things and give you more facts about the food truck industry. Read on.

Chapter 1: Top Reasons to Open a Food Truck Business

There are distinct advantages that food trucks offer when compared to restaurants or traditional foodservice businesses. The average restaurant manages to load its owners with overhead costs and other expenses before they even begin. Running a restaurant is a serious task, and if you've never run a business before, it's close to impossible to succeed.

Lower Costs

This is not the case with a food truck. Your biggest expense is going to be the truck itself. On average, a food truck's startup costs will run you anywhere from $50,000–$60,000 (25 Powerful Food Truck Industry Statistics in 2020, 2020). This is much lower than what restaurant owners face. With a restaurant, let's look at the number of things you need to purchase before opening your doors.

First, you need tons of kitchen equipment. You'll be serving customers who expect fresh food, and you'll presumably be renting an ample space. This means you'll need top-notch equipment that can store and cook food. You'll need to hire at least one more cook in the kitchen since you can't do everything by yourself unless you want to go mad. Someone will need to wash the dishes and clean the floors constantly. Your restaurant will need to be open for long hours since you'll need to cover more tables to make your money back. Look at the life of an up-and-coming chef, and you'll see how tough it is. A kitchen is a demanding place, and you need to remain in it for long hours to have any chance of success.

Of course, none of these deals with the front-of-house issues. You need an able supervisor who can marshal your waiters and employees to ensure good service. Finding such people is hard. What happens if they quit or do a bad job? You can cook the best possible food, but if none of it reaches customers on time, they aren't coming back.

A restaurant's success depends on a lot more than just its food. The added costs increase pressure on owners to focus on many things other than the food. If the restaurant gets its cuisine choice wrong, they'll need to restart all over again, which means more expenses. With a food truck, you simply change the branding, and that's it. You don't need to worry about waiters or cleaning tables. Thanks to lower costs, your break-even points are lower, which means lesser hours spent working. You can fix your hours better and keep more of the money you earn. What's not to love?

Brand Recognition

It turns out that food trucks are also a great way to open a restaurant in the future. What is it about all great restaurants have? Brand recognition. If you walk down a street full of restaurants, chances are you'll pick the one you recognize or feel an affinity for. Choice of cuisine plays into this, of course, but people will go with the brands they recognize for the most part. It's why Nike shoes sell for five times the price of their cheaper competitors.

To build a brand as a restaurant, you need to invest a lot of money in advertising while juggling the other things I highlighted before. This task is well beyond most restaurant owners. Famous restaurants capitalize on the popularity of their chefs and, on the way, they cook their food. Some try to make a splash through decor and other innovative concepts. If you are trying to bootstrap their business, opening a restaurant is not the way to go.

Instead of splashing all that cash on a fixed location, start your business with a food truck instead, and build a reputation in the minds of your customers. The food truck scene is always lively and is well covered by local media since there are so many choices available. A food truck serving Greek food last week could be serving Thai street food the next. Consumers are always kept on their toes, and dedicated food truck parks seek to keep the food on offer as varied and exciting as possible.

All of this means you get free publicity without having to do any work. I'm not saying you don't need to publicize your business or invest in advertising. However, you certainly don't need to advertise and invest as much as a restaurant owner would. A roving food truck also generates excitement. Food truck locator apps exist for a reason. Consumers feel as if they're receiving a present of great food when a famous food truck visits their neighborhood.

Call it the ice-cream truck effect or whatever; food trucks generate excitement and buzz when done right. This means you're building a good customer base for the future. If you open a restaurant down the road, it will probably be packed. What's more, driving your truck to different locations and selling food is a great way to get to know a place and test demand in that area. Restaurant owners typically spend time in an area before opening a site. This costs them money since they aren't selling anything.

As a food truck owner, you can park, observe, and earn, all at once. Another advantage is that you'll be running a restaurant without all the pressure that comes with running one. You cook great food and leave it at that. You don't need to worry about dodgy waiters and other factors that will ruin your customers' experience. You'll get to know how a particular location's legal system operates and can make a better decision whether you want to open a restaurant there or not.

Passive Income

The restaurant business is tough, and even successful restaurants require their owners to be present at all times. You cannot outsource the running of a restaurant to a manager without giving up a lot of control and paying them high compensation. Look at how franchise restaurant owners spend their time, and you'll notice that their lives revolve around their businesses, leaving them with very little time to do anything else.

Some people might be happy with this, but I'm guessing this sounds unappealing if you're like most people. A food truck requires work, and you'll need to put in the hours. However, these hours are lower, as I mentioned earlier. The other excellent quality of a food truck is that there isn't much to manage, just a few items. You need to maintain your vehicle, ensure you're abiding by local laws, and serve great food. That's a lot less compared to a restaurant.

This means you can realistically hire a manager and gradually give them more responsibility as you step back from the business. Once you've stepped back enough and have amassed enough capital, you can franchise your business and earn royalties on it. You'll find that doing this is much easier with a food truck than with a restaurant.

A restaurant franchise will make you more money undoubtedly. However, your chances of success are lower with a restaurant. Besides, you'll face the same issues when you open a new location. You'll need far deeper pockets than what you'll need for a food truck. This makes it easier to attract franchisees and earn royalties. Passive income adds up over time, and soon you'll be generating a high income without having to spend time maintaining it.

Chapter 2: Governing Law

Taking your food truck out on the road is exciting, but before you begin preparing and selling food, you need to understand the important laws and regulations surrounding food trucks. You need to have the proper licenses, insurance for your business, and you will need to follow important health and safety guidelines. Certain taxes will need to be paid as well. Dealing with all the laws, licenses, and regulations can be one of the most frustrating parts of starting your street food business, but with some help, you will get through this step and on to indulge in your passion for food.

Taxes

Since you are self-employed, you will have certain taxes that you need to pay. Tax law requires that you register with HM Revenue and Customs within the first three months of being self-employed. Self-employed individuals are responsible for taking care of their taxes and their National Insurance contributions. Each year, you will be required to fill in a tax return. If you are not sure what you need to do to file taxes when you are self-employed, you can call the HMRC helpline or talk to a local tax office.

Food Licenses and Registration

To begin operating your street food business, you will need to have the proper food licenses and registration. Food businesses must have the necessary registration with their local council before starting to trade. You will need to register your street food business with the local council in the area where you store your vehicle at least 28 days before opening your new business. Do not make the mistake of registering where you operate—make sure you register with the local council in the area where your vehicle is stored since registrations are checked regularly.

Suppose you are running a mobile food premise. In that case, you are also required to have a valid vehicle excise license (which is a road tax), an insurance certificate for your vehicle, and a certificate of roadworthiness for your car. Individuals who will contact food also have to have appropriate training. The training level required will depend on the particular job. It is also possible that you will need some specific licenses, depending on your business. For example, if you sell alcohol, you will need support for liquor and entertainment. This license is also required if you plan to sell beverages or food between 11 p.m. and 5 a.m.

Insurance

To legally run your street food business, you need to be properly insured as well. Whether you are using a trailer or a van for your business, you need to have good insurance. Several insurance options are available to you when you start looking for your insurance.

Product Liability Insurance

This insurance is a great idea to run a street food business in the UK. This insurance will help keep your business protected from claims made against your company if beverages or drinks are accidentally served that cause food poisoning. Medical expenses will be taken care of for those affected if the request is justified.

Public Liability Insurance

This type of insurance helps to protect your business from claims the public may make against you due to accidental injury or accidental property damage that occurs at the hands of your street food business.

Employers' Liability Insurance

If you hire employees, you must legally carry employers' liability insurance. If a member of your staff is made ill or injured accidentally on the job, this insurance covers damage pay-outs and legal fees that you may face. In a hot kitchen, accidents can quickly occur, so this insurance is essential.

Catering Van Insurance

Not only do you need insurance for your business, but your van must legally be insured as well. Many companies offer catering van insurance, with various levels of cover available. Of course, while different policies can offer many additional features, the following are a few essential features you may want to consider.

No Claims Bonus

Suppose you already have a no claims bonus on your private vehicle policy. In that case, insurance companies may be willing to offer you a no-claims bonus on your catering van policy as well. Ask your insurance company about this feature before choosing your insurance.

Comprehensive Cover

If something happens to your van, your business could quickly go under. This means you will probably want to consider a comprehensive cover for your van. Ask about new for old cover as well, so if your van is a total loss, you will receive a pay-out that considers what you will have to spend to get a new van for your street food business.

Road Assistance and Towing

You cannot afford to be left high and dry if you break down. When purchasing the insurance for your van, look for a policy that offers roadside assistance and towing. Breakdown cover can be a great option, which can help you recover the loss you experience if a breakdown occurs.

Contents Insurance

A general insurance plan for your van only covers the vehicle. Since you probably will be spending a large amount of money on equipment in your van, having contents insurance is a great idea. This way, you get coverage for your fryers, stoves, fridges, and other equipment you have in your food truck.

Health and Safety

Certain health and safety regulations must be followed. If your food truck does not follow these regulations for mobile food businesses, it could affect your right to trade, which your local council enforced and administered. You may have to undergo food safety inspections at regular intervals. Any waste from your mobile business must be disposed of as trade waste, meaning that you cannot dispose of it as domestic waste.

Cross-contamination is one of the big risks that mobile catering businesses face. To avoid this, certain health and safety guidelines must be followed. Some of those guidelines include the following:

- Cleaning out sinks after raw foods or vegetables are washed or prepared in them.

- Avoid touching food with your hands.

- Chopping boards and food preparation areas must be cleaned and disinfected.

- Cooked food and raw foods must be kept separate.

- Food must never be stored on the ground and should be at least 45 cm above the ground.

- Equipment needs to be disinfected and cleaned after every use.

Your food truck must also follow certain temperature control and storage regulations. These guidelines are important because it helps to avoid food poisoning. Some of the policies you should be tracking include:

- Cooked food needs to reach 75ºC at its core.

- Hot food should be kept at a minimum of 63ºC.

- Food should be immediately prepared before serving.

- High-risk foods like dairy and meat products should be kept cold at 8ºC or below.

- Thermometers should be used to check the temperature of cooked foods.

- When storing food in the refrigerator, food should be covered.

Of course, these are just a few of the health and safety guidelines and regulations that must be followed. To learn more about your area's food safety requirements, contact your local councils' Food, Healthy & Safety division for more information.

Chapter 3: How to Start Your Business

The meals truck enterprise is one of the fastest-growing industries for a purpose. The company is smooth for marketers to recognize, at an equal time, as clients cannot get enough food vans from their towns. So, now that you see how popular food cars may be, how can any properly—which means person get started with their very personal meals truck business?

To begin, even though starting a food truck is considered simpler than starting a restaurant, it's nevertheless hard to get off the floor if you're now not prepared to place within the effort to make it work. Meals trucks can become exceedingly at risk of succumbing to the factors and be difficult to many risky outdoor factors like special cars, theft, vandalism, and even sudden fire dangers.

Running a food truck during certain seasons is grounds for now not getting any income. I advise you not to run your food truck in the winter while promoting mostly frozen treats like ice cream. The hours can be long; some food vehicles stay out from 9 a.m.–5 p.m. on a regular schedule, even as others turn out to be spending the entire day and night time in one place.

Regardless of those genuine troubles, beginning a food truck is a great way to start as an entrepreneur in a familiar location. If you are obsessed with the idea of cooking, possibly starting a food truck can be an excellent business for you.

Questions to Ask Yourself Before Opening a Food Truck

Starting up any business takes time. Even though, so long as you make an effort to get a business up and running, you should not be afraid of capacity faults that can arise when running the business itself. This notion also applies to people who are counting on getting into the food truck industry. However, they do not anticipate getting involved in the business until they are about to make contributions.

However, before you determine what form of meals truck you are planning to open, it's essential to invite yourself questions about what you could assume out of business. Those questions may include:

- How is my passion for food?

- What type of food do I want to serve?

- What kind of food may be missing from my community?

- Are there other food trucks with the same concepts/food products as mine?

- How can I differentiate my brand from different brands?

- How much money will I have to invest in this business initially?

- Where will I get the food that I'm planning to serve in my business?

- How much do I need to invest in those products?

Asking allows you to postpone the uncertainty of beginning a food truck enterprise. In fact, for numerous people, understanding greater approximately what to expect from a meals truck business makes them more relaxed with beginning the enterprise inside the first location. They do know what to anticipate after doing the studies.

Once you analyze what to anticipate, there are different elements that you have to take into consideration, especially considering starting a food truck enterprise. The factors are a crucial part of the company itself, so it is very important to note them.

The Most Integral Factors of Starting a Food Truck Business

The essential factors of beginning a food truck company are pretty straightforward when you remember. You need to take care of these issues before you hit the road for the first time! The intensity that these elements address will be reviewed below:

Legalities

You can't run a food truck business without seeking some form of a permit from your city and country's jurisdiction. Having a permit is certainly specially designed to cover up a mobile agency, take care of the legalities associated with the company, and the automobile's operation inside your city.

At remarkable, a permit safely covers strolling most elements of your mobile food truck; besides, you want greater criminal permissions from your city or county jurisdiction.

Vehicles

You want a good truck for your business and success in this venture. Many food trucks and/or trailers can cost anywhere from $1,500–$75,000, depending on the type of truck you may plan to purchase.

In the end, you need miles large enough to house the business and stay for some time.

Logo

A brand allows potential customers to understand a business, especially if that business is already fairly well known. As a food truck owner, you will want customers to remember the idea of your business and the food they eat, so start brainstorming names and thoughts that can further what you imagine your business to be.

Names are also just one part of setting up a business; you have to create a menu, decide what dishes to cook, and put together everything you need before you even get the legal paperwork together.

Financing

It is arguably the most important issue when it comes to starting up a food truck. Why? Due to the fact you need some funding. So, before you get into the question of planning, the search for financial capability must begin. You might have the desirable fortune of having a private investor who may be interested in backing your food truck business, even though it rarely happens that an investor takes the idea to shape what you want. So you could get a loan from a financial institution, specifically in case you qualify for any of their financing options. You could seek help from a monetary corporation if you do not want to apply at a bank.

Financing a food truck normally consists of protective expenses, branding, system, food, associated additives, point-of-sale (POS) or credit score card systems, protection precautions, and employees.

Therefore, there are several factors to consider for starting your food truck business. So you need to review these factors so that the business's success is guaranteed.

Chapter 4: The Primary Causes That Lead You to Bankruptcy in Food Truck Business

I've listed the advantages, but there are some pitfalls you need to be aware of. I don't mean to present food truck ownership as the key to great success. The truth is that it takes a lot of work. You'll need to execute many processes well and constantly evaluate how your business is doing. Food truck owners get lazy after achieving initial success, proving to be their downfall. Here are some of the other pitfalls to beware of.

Lack of Planning

There are a few items you'll need to plan before you get your business up and running. Chief among these is your business plan. Your business plan outlines everything relevant to your food truck operation. It lists what your locations will look like, what kinds of customers you'll target, and other financial information.

Most importantly, it'll also outline when you can expect to make your money back.

This is a crucial point to consider when starting any business. You cannot run a business without capital, and too many business owners fail to consider their true costs. They assume their ventures will be successful from the first day, but this is never the case. It takes around six months to a year for your cash flow to stabilize. You need to account for these low sales periods and have enough cash in the bank to pay your bills and maintain your standards.

Without proper planning, you're unlikely to be able to project any of this. Planning is tedious, and it isn't exciting. But I'm going to show you how to turn this thought process on its head and make planning something you look forward to. Always remember: If you fail to plan, you plan to fail.

Doing Everything Yourself

Most business owners tend to love having control. However, most business owners also fail. The successful ones learn to let go of their need to control everything and delegate and provide clear instructions to their employees and the people assisting them in their business. The average person has a warped view of the tasks a business owner carries out.

Common wisdom says that a business owner needs to wear many hats. They need to understand their business, understand operations, understand marketing, figure out taxes and accounting, and make time to pay themselves. The mistake occurs when the common person thinks that a business owner needs to execute all of these things by themselves. There's a difference between wearing a hat and being a full-blown professional at everything.

As a business owner, you need to understand whatever is relevant to your business in each of those fields. You're not a superhero and aren't perfect, no matter what your dog thinks of you. You cannot hope to become an expert in all of those areas. Instead, you need to focus on what you're good at and consult with those who know what they're talking about when you need to deal with the other stuff.

For example, suppose you're seriously considering opening a food truck business (since you're reading this book, I'm assuming you are). In that case, you're probably well versed in cooking food and understand how to present your cuisine. Let's assume you don't know a thing about marketing, about legalities, about accounting, about setting up a business structure, or even how to run a kitchen. Here's the good news: You don't need to know all of this.

It would help if you learned whatever you could and outsourced the tasks that you could not comprehend. Hire an accountant and use their services to open your business and file your taxes. Hire a junior cook to help you in the kitchen, talk to other food truck owners, or observe how they run their operations. If you're deficient in marketing, hire a great graphic designer to create logos and marketing material for you. Have someone make an excellent website for you, and then hire a freelancer to incorporate a location tracker.

Whatever you need, there's help for it. The internet has made it so easy for you to find the right people that all you need to do is click a button. So let go of the need to understand everything about everything. Even Michelin-starred chefs hire pastry chefs to prepare desserts. Do you think they sit down and prepare their accounts themselves? Do they fix their plumbing issues themselves? They can't even prepare their food for themselves; they need an army of chefs to do so. So let go of the need to control everything, and instead, focus on your strengths. Outsource the rest, and success will follow.

Poor Service

You won't have to deal with front-of-house operations when running a food truck, but this doesn't mean you can ignore customer service. To be a successful business, you need to listen to your customers and give them what they want. This is a tough pill to swallow for many restaurants and food service business owners. Why is this? Let's call it the artistic temperament.

Everyone has ideas of what food should taste like and what works best. You might think that salt is overrated and might refuse to add any to your food. However, if most of your customers prefer tons of salt in their food, who are you to argue? A famous chef can get away with kicking out customers who want to change the way their food is cooked, but you can't.

This doesn't mean you need to compromise your ideals or serve unhealthy food. However, you need to strike a balance between your artistic side and the health of your business. Compromise is necessary, and you need to remind yourself of why you're running a business. Be passionate about the food you cook and be equally passionate about running a good business. You're probably not the only one who depends on its health.

There are no tables to clean or water to serve, so this makes customer service quite simple. The typical food truck customer wants food quickly and served fresh and tasty. They want their food cooked right and served with a smile. These aren't tough things to do for any business owner. Have fun running your business, and your customers will have a great experience.

Don't compromise on the quality of your food or you're cooking to make more money. In the short-term, you will make money, but your customers will recognize the drop in quality and visit some other business in the long term. Make sure your food is as fresh and delicious. To save time, many food truck owners (and restaurants) cook food beforehand and freeze it. This robs food of its texture and taste. Reheating food in the microwave is no one's idea of cooking.

This is where planning comes in handy. If your menu is elaborate and is full of food that cannot be reasonably prepped beforehand and cooked at short notice, you'll need to change your menu. A food truck customer isn't going to wait for more than five minutes for their food. The idea is to deliver great taste quickly and conveniently. Tailor your menu to achieve this, and your customers will always be happy.

Ignoring Budgeting

This is a mistake you think most business owners would avoid, but it's alarmingly not the case. Many business owners fail to keep track of their receipts and merely guess how much money they're making. It's usually a loss, so; there's nothing to report there. If you don't keep track of spending and budgeting in your personal life, then forget about opening a business right now.

You're not going to start budgeting once you open a business magically.

You'll only replicate what you do in your personal life. If this involves zero tracking or very loose tracking, you're headed for trouble. A business isn't a game that you should take lightly. You need to be on top of everything. This isn't as hard as it seems, but it requires you to establish processes and consistently practice them.

For example, it should be standard practice for you to deposit the day's take into your business bank account at the end of the day. Much like how kitchens need to be cleaned after the end of the day's service, you also need to perform some financial tasks. At the end of every month, you need to take stock of what you have done and plan. You might need to invest more in improvements or marketing to grow your business.

Cash flow is a challenge for every business owner. You never know when a crisis might occur, either in your own life or with the general economy, and you need to prepare buffers against these. Look at how badly small businesses have been wrecked thanks to the COVID-19 pandemic. It exposed just how unprepared most business owners were. I'm not saying they ought to have predicted how bad the pandemic would get. But they should have held reserves of cash to prepare for the unexpected.

Just as you save for a rainy day, your business needs emergency cash on hand as well. The rule of thumb in personal finance is to keep six months' worth of expenses as cash. Carrying this over to your small business accounts is an intelligent move and will give you a reasonable margin of safety. Make tracking finances a priority if it isn't already one. Also, make sure you incorporate good habits into your personal life. Carry these over to your business, and you'll avoid falling into a debt hole.

Poor Marketing

You cannot rely on the old marketing adage "Build it, and they will come." That's not how marketing works. You need to advertise and target your customers consciously. These days, thanks to advances in digital marketing, it's possible to laser-target your customers.

You need to know them beforehand (not personally, but their habits) and target them accordingly.

Any business that still relies solely on advertising in print media and hopes word-of-mouth spreads is preparing to fail. Your social media strategy needs to be robust. You don't need to keep posting every hour of the day, but you do need to remain active. If social media posting scares you, consider hiring someone to manage your profiles. Many business owners do this, and it's an example of how you can outsource the tasks that you don't enjoy.

Marketing is all-important these days because many things compete for your customer's attention. You don't need to scream and shout to be heard, but you do need to invest in marketing and brand creation. Many marketing terms will sound like nonsense to you, and marketers can indeed take themselves a little too seriously. This doesn't mean they're wrong, though. Educate yourself on marketing basics and invest in good branding materials. They'll help you stand out more.

If any of these worries you, then just keep reading. You'll gain a good grounding in all the necessary topics by the end of this book.

Chapter 5: The Importance of Social Networks

In addition to running the day-to-day operations, it's important to understand just how crucial social media is in today's market and how it affects your own business. Rather than feeling lost in the social marketing world, you need to start early to build a vast social media following and get familiar with the process if you aren't already. Numerous food truck owners still don't understand the importance of social media for their mobile businesses.

For food trucks, it's important to be active on Facebook, Twitter, and Foursquare. In addition, it's also a good idea to consider Yelp and Instagram. Equally important is having a website for your food truck that works as the main hub of all your social media interactions.

At first, it may seem daunting to have so many social media responsibilities. However, when done correctly, you will enjoy the benefits of free marketing and advertising. This can and will save you lots of money from doing traditional marketing and advertising. So be sure to post updates frequently, allowing new customers to find you while keeping loyal customers paying customers.

Being Unique on Social Media

There are countless other ways to market and advertise your food truck. Due to the popularity of food trucks nationwide, these mobile businesses have had to get creative to stand out from the crowd. *Mexicue*, for example, invites outsiders to post recipes on their Facebook page so others can vote on their favorites, awarding the winner a menu spot, along with several gift cards and promotional gifts.

Ninja Snowballs, a Baton Rouge snow cone truck, felt the decline in business during the colder months of September and October. So, in the fall months, the food truck decided to host a flavor contest, bringing an additional 1,500 views to their page and keeping the business afloat during those cold months.

These are just a couple of suggestions to help new owners get their minds working on social media. Staying in touch and keeping open communication with followers is perhaps the best way to manage social media followers. Using social media is as important as having a friendly face at the service window. Bring out your personality and let customers know you as a friend and business mobile.

Connecting with Your Customers

Once you start getting followers on your social media accounts, you need to monitor the activity on those accounts. This is sometimes the only way that you get feedback from your customers. It's where you can make announcements and share other news from your food truck without being a nuisance to your followers.

It's important to interact with your followers on social media to seem approachable. You're also building a relationship with your current and future customers. There are enough tasks for you to do just running your truck, so you don't have to feel like you need to respond to every message (although some do), but being semi-active on your accounts makes it worthwhile for people to follow you.

Social media can also generate more customers and reward those you already have. Discounts, coupons, special events, and other promotional announcement can and should be made through social media. It's often the best way to make last-minute announcements. It's also an effective way to let your customers know if you can't make it to a regular location. Social media can reach many customers quickly and effectively, so make sure you use it to your advantage.

Chapter 6: Advanced Tips for Using Social

Luckily, we live in an era where many cost-effective options for marketing a food truck. Because of technological advances, advanced marketing techniques are easily practiced by all. Here, you'll learn about the all-important concepts of marketing and promoting your business. This will directly affect your brand, visibility, and eventually your income. We already know that good food and great customer service are the foundations of your business. But that's not enough! You won't be selling much food if customers don't know your business exists.

Marketing and promotion are vital components in any successful business. And it's important to have your marketing plan developed early on. Often, the first step in marketing is choosing a name for your business. In this case, it is your food truck. The food truck business is very competitive and picking a good name is the key.

There are plenty of good and memorable names out there. But it's up to your creativity to develop a good one that fits your business.

Picking a good name can be difficult. That's when you can ask for help to find a clever name. You could turn to friends and family or hold a contest. Ask your followers on Facebook and Twitter, or use your blog to ask for suggestions. However, if you use social media to find a name, just be sure to offer a reward to the person who comes up with the winning name.

All these elements will contribute to the overall branding of your food truck. Once you get rolling, you can consider putting your designs on cups, napkins, coasters, business cards, and other stationery. Your designs can create an overall mood and atmosphere around your brand.

Once you have your logo and branding, it's time to move to social media to help promote your business. Just about every mobile food business use social media to market their truck. The top two services that truck owners use are Twitter and Facebook.

Tweeting Your Way to Success

If you're new to Twitter, here's what it is. It's a social networking and micro-blogging platform that allows up to 140 character-status updates. Twitter is free to use, and the more followers you have, the further your messages reach. Twitter provides real-time updates for you and your customers.

You'll be using Twitter to announce your location, specials, new items, and other important messages. Twitter can also support two-way communication so you can respond to your followers through direct messages. It's also good practice to respond to Tweets directed to you to keep your customers engaged. However, it's not necessary to respond to all tweets. But giving some response will show that you care about your customers.

When you're getting started with Twitter, you'll need a username, e-mail address and come up with a password. For people to recognize you, come up with a creative username. This is usually the name of your truck. It might be a good idea to search Twitter to see if the username you want is available. You should do this when you are first coming up with names for your truck. Once you have a Twitter account set up, you'll want to upload a profile image. It's a good idea to use your logo or a photo of your truck.

Facebook Status Updates

The other popular social network that trucks owners are using is Facebook. Those who don't know what Facebook is are social networking platform that allows status updates through text, images, and video. Unlike Twitter, Facebook doesn't have a 140-character limit when typing updates. But similar to Twitter, Facebook is free to use. The goal on Facebook is to get "Likes." Getting "Likes" is similar to gaining followers. And the people who have "Liked" your Facebook page will instantly get all your updates in their account.

To use Facebook for your business, you need a business page or a fan page. If you already have a personal Facebook page, you can easily create a fan page. However, we don't recommend using your page for your business status updates. It's a good idea to keep your updates separate from your professional business. Once you're all set up, start using Facebook to update your locations, menu items, share photos, videos, and more!

Whenever you can, encourage people to "Like" your page. When people "Like" your page, they are essentially subscribed to all your updates. Updating your Facebook status frequently is important for engagement. Once your social media accounts are set up, you'll want to promote your Facebook and Twitter profiles everywhere! This information could be printed on your truck graphics, business cards, and website.

How to Start Your Website

The next step in promoting your business is to create a website. Having a website is equally as important as using social media. A website is essential for any business, whether it's online or offline. A good web designer or developer can develop a complete website for you. It would help if you viewed sample websites from the web designer first before hiring them with logo designers. Sometimes the terms web designer and web developer can be used interchangeably.

To start a website, you will need a domain name and a hosting account. Your domain name is your website address. An example would be www.yourfoodtruckbusiness.com. A hosting account is where your website files are stored and accessed by visitors, kind of like a remote hard drive or computer. But let's take a step back before building your site. First, you need to see if your domain name is available. Again, it is helpful to come up with a domain name when you are picking a business name. But once you've decided on your domain name, you'll need a hosting provider such as Bluehost. Bluehost gives you a free domain name that you own when you host your website with them.

But here's a website-building tip that can save you a lot of money. Don't pay a web designer or developer to build your website! You can build the website yourself without any experience! It's easier than you think! You can simply buy hosting and the domain yourself. Then install WordPress to manage your website. You may or may not know what WordPress is, but it's a trendy blogging platform that's free and easy to use.

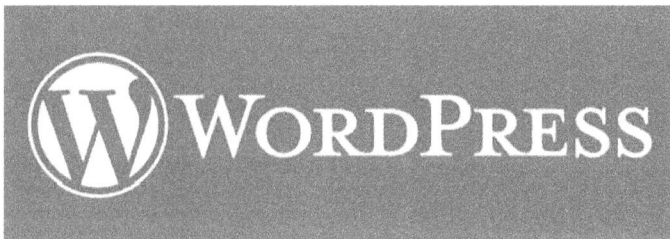

Many food truck websites use WordPress as their preferred management platform. The best thing about using WordPress is the free themes you can use to customize the look of your site. With so many themes available, you can change the look of your site instantly without the help of a web designer. With WordPress, you can easily create any pages yourself, such as the

Homepage, About Us, Contact Us, and Menu pages. Let's not forget the original function of WordPress, which is its incredible blog capabilities.

Even if you don't choose to create your website yourself, a designer would probably use WordPress anyway! So why pay them for something you can do on your own! At the very least, buy your hosting and domain name yourself before turning it over to a web designer or developer. They'll probably charge you a set-up fee for the initial work with your hosting account! You can easily set up WordPress yourself through the hosting control panel. Almost every hosting provider makes it easy with one-click WordPress installs. So why pay a designer for something you can do yourself? People with literally no web building experience have built beautiful websites with WordPress.

This is just an overview of some of the marketing strategies that need to be employed by food truck owners. Suppose you want a complete food truck marketing plan. In that case, the next book in this series, called The Food Truck Marketing Handbook, explains in detail the tools and effective strategies that will help build a strong social media following, including tactics not found anywhere else. It is completely tailored to the food truck owner.

Chapter 7: Successful Marketing Practical Ideas for Your Business

When you are ready to promote your great food to a hungry market, you must tell the world that you are ready for business. Give people reasons why they should come! Focus on all the factors that encourage a new customer to find your location and view you as a favorable option. You want them to stop and buy food from you, and most importantly, come back and buy from you repeatedly. This is the formula for success for any product at any location.

You will not be able to sell your food to everyone! Focus instead on the most profitable market and spend all efforts reaching and securing it first. Then spend extra time and money going after new groups of customers to expand your business. Here are some questions to consider during your marketing efforts:

- How can you stand out in the market?

- What niche does no competitor fill?

- How can you be unique and recognizable?

- Can you specialize and focus more?

- How are your competitors positioning themselves in the marketplace?

- What is their advertising strategy?

- What benefits are they offering to the consumer?

- Can you offer something that your competition doesn't?

- Are you looking at all segments of the market?

- Larger companies often fail to realize small segments that could be perfect for you!

- Who is your customer?

- How can you reach him?

- How can you reach out and target other markets?

View Your Business Through Your Customer's Eyes!

You should take the necessary time to understand as much as possible about your local market. How can you make good decisions about marketing your operation if you don't consider the potential of all customer groups in your area? You will have to research what they want from you!

All people get hungry and eat regularly, but that knowledge alone will not assure your success. Why not? Because many factors influence how people see your business and help them decide if you offer an attractive option.

Remember the last time you were at a fair or any other large public festival? You probably saw many foods and drink options lined up side by side, all anxiously trying to grab as many customers as possible. After all, each one of them probably paid a lot of money for the opportunity to be there. Each vendor was also under pressure to earn back the investment and make a decent profit at the event. Now, if you took the time to notice, I can assure you that not all vendors were equally busy.

Some vendors were drawing a larger crowd than others. Very few vendors were extremely busy and had long lines of people in front of the sales area. Each of these vendors arrived with the idea that they would attract many customers, but some did much better than others. Why? They understood what customers wanted, and they provided it! I am not just talking about food here. I mean the whole package of things that make a customer buy like:

- Location.

- Price.

- Signage.

- Ambiance.

- Attractiveness.

- Consistency.

- The attitude of employees.

- Showmanship.

- Entertainment.

- Efficiency.

All these factors work together to help attract customers and make them buy. Not understanding these important factors will relegate your operation to one of the slow businesses that are lucky to break even, much less turn a profit.

Target Your Customers!

There are many customers, and each group wants something different from you. You will be most successful if you first identify what each group wants. Then you can focus on providing their wants and needs for them. You have several options to target the most profitable customer groups in your area. You can target them by price, menu, location, or your business goal. Look at the following questions. Maybe they can point you in the right direction:

- Do people at construction sites, offices, residential areas, etc., have to travel too far to get a convenient meal? Target these groups with your location!

- Are there some popular food items you are sure would be very popular in the area and are not offered by others? Target with your menu!

- Is all the competition in the area too expensive for the average population? Target by price!

- Is your business goal to only work a few hours per day from Monday through Friday while your kids are in school? Find a location that will provide you with a busy lunch business and make your money fast!

All these are powerful questions to ask. If you answer them, you will create just the kind of business that meets your needs. The power to control your time and income is one of the primary reasons to be self-employed! Now that I've given you some ideas let me tell you about the process I use to evaluate a market and all the information I collect, so I can make an educated guess as to how successful I might be in a particular location.

Let the Customers Guide You!

Look around in your town for target areas. Students and faculty staff of universities, colleges, and schools are often underserved and offer excellent growth potential. If you can tap into tourist areas, you will have a steady stream of hungry customers, always on the lookout for local foods. To target this customer base, you should also look into expanding into other items. Maps, postcards, and souvenirs should all have their place in your sales window if people ask for them and are ready to buy!

Listen to customers' questions and never hesitate to ask if there is something they want to see on your menu. You will be surprised about how many ideas you can integrate into your operation. A vendor I talked to started offering a "Dog-wich" for customers' faithful dogs that were waiting in the cars! He sold them at an impulse price ($.99), used up some of the lesser quality meats, and won many loyal followers. It all started with a customer's comment and the action of a creative vendor. Yet another moneymaking idea was born!

In good locations, about 60–80% of people that purchase from you will be repeat customers. You need to take good care of them, or somebody else will! Get into the habit of putting on a friendly face when dealing with the public. Leave them with an experience of satisfaction, and you will be sure to have them come back! Many vendors do not understand the strategy of repeat sales. They are just looking for the maximum profit in each transaction. Customers will notice this attitude and will eventually start feeling ripped off.

Creative Advertising Ideas

Try out something new and consider the following ideas:

- Purchase small ads in the local newspaper to let people know where they can find you. Tell them about lunch specials and plant the idea in their head to take food home for dinner.

- Give out refrigerator magnets with a fun logo, quote, or saying on them. This way, people think of you when they think of food!

- Drop off menus with coupons in all surrounding businesses, offices, neighborhoods, and potential wholesale purchasers.

- Send out a brochure (direct mail) to businesses selling automobiles, mobile, or manufactured homes and offer to set up for sales events on weekends.

- Look for wholesale business! Talk to restaurants and see if they would be interested in purchasing food from you.

- Consider a short radio ad in a local radio station.

- Have a postcard designed and printed with a picture of your best setup. Make sure to have some people in the pictures as well! Have these postcards on display at your counter for all to take. Some people are in love with beautifully printed paper. They will hang them everywhere, and many more people might see them. They might also send them to someone.

- Have a sign offering free samples! Everybody likes to get something free! Get them to taste, and you have them hooked! This way, you will attract many new customers to your business!

Marketing Campaigns

Any concession trailer business will see significant results from menus or coupons given out every order. This strategy even works at weekend shows and festivals, as you have a chance of convincing customers to return a second time or send their friends. You should always give them a reason to come back and choose you over the competition. Happy customers are like money in the bank! They will make you rich over time! Here are some tricks I have learned:

Coupon Magic

One of the best ways to bring new and old customers back to your business is coupons! Studies have discovered that the offer inside the coupon does not grab people's attention! Neither is the price! Test runs with different prices brought similar responses. Even coupons containing the usual price are often used! It is simply the dotted line around the coupon that will make people take note and get ready to grab their scissors! This is why it is also a good idea to add a symbol or graphic of a pair of scissors. Coupon collectors will save any that suggests value. Coupons work wonders, and you should make them work for you! You can do several things to make coupons work even better:

Add a Free Bonus!

Your bonus item can be anything as long as it has a perceived value. Better yet, give a second item: "FREE BONUS Sandwich with the purchase of a Sandwich!" This helps to bring several people in. You can also add the bonus to your slow-selling menu items with high-profit margins. This tempts people to spend a little more because they want the "bonus." Think about all the options you have in your operation to add bonuses to your coupons! Add your bonus to bold letters so it stands out! Capitalize some of the keywords like "FREE" and "ONLY" in your coupons as well:

The Right Hook

Oftentimes it is a bad idea to give out coupons that state something like "$2 off your next order" or "$1 off all large Sandwiches." These discounts tend to cheapen the perceived value of your products in the eyes of your customers. When you have several coupons floating around with different discounts, this matter gets worse!

Discounts raise suspicions. People might feel that your products could be overpriced to start with or even worse: cheap! In the worst-case scenario, people on the road will not visit you today, as they left their coupon at home! The discount coupons do something else that can backfire: They will train every person that ever used one at your business to wait for your coupons to land in their hand before they even think about stopping by again. You are targeting only a small area of your potential customers, the "Price Shopping Bargain Hunters." Instead, you should bundle complementary menu items together to give "added value" to your offer. Now you are targeting people attracted by a "good deal" rather than a "cheap price"! Here is an example:

BONUS DEAL
AT SMOKY MOUNTAIN BBQ
TWO
JUMBO BBQ SANDWICHES
FOR THE PRICE OF
ONE

Keep Track of Your Efforts!

Collect all coupons and find out which ones frequently come back to you. These are your winners. Try out different offers and see if you can hit an even better return rate! Keep track of the losers as well! Do not use coupons that fail to do the job! This rule equally applies to print advertising! If ads don't bring the expected results, pull them immediately! Those ads just drain your bank account! It is not enough for an ad to just "keep your name in front of people's eyes"—it has to bring people to your sales window actively! That is why you should always give your coupons a week-long expiration date! Lead them to the action! If you try out different designs and keep track of the results, you will end up with several winning coupons and ads that will draw people like magic!

Effective Advertising

Newspaper ads in local magazines and newspapers are another excellent way to increase your business. One mistake that many vendors make in their advertisements is based on the product they are selling. This way, they are missing one of the best advertising benefits: advertisement for their business name!

Giving your menu items signature names helps you build unique products that customers can only buy from you! This technique can dramatically increase sales, as people are attracted to unique offerings! Interesting names draw attention! Keep this also in mind when you design your menu! It should have your company (brand) name featured as many times as possible! Here are some words that can entice people to buy like crazy!

- Popular.
- Top Choice.
- Fun.
- Market.
- Fresh.
- Famous.
- Original.
- Favorite.
- Award-winning.
- Natural.
- New.
- Gourmet.
- Imported.
- Featured.
- Old fashioned recipe.
- Quality.
- Original.
- One of a kind.
- Featured.
- Garden taste.
- Grilled.
- Roasted.
- Ice cold.
- Aged.
- Light.
- Extra.
- Special.
- Creamy.
- Savory.
- Marinated.
- Smoked.
- Giant.
- Steamed.
- Healthy.
- Custom.
- Stuffed.
- Melted Cheese.
- Toasted.
- Jumbo-sized.
- Ultimate.
- Unique.

You can play around with different combinations of these words. Use them liberally in your advertising flyers and throughout your menu! How about: "Try our giant, famous, grilled, gourmet, garden-fresh sandwich with toasted almonds and melted cheese served with our special featured fire-roasted pepper sauce."

Don't Attract the Wrong Crowd!

Another mistake many operators make in their advertisements and coupons is to compete by constantly discounting their prices. This tends to be a downward spiral, as price alone is a very poor reason to be selected to start with! It attracts the "Price Shoppers" that don't care about you or your product. All they care about is the lowest price! As soon as they find it somewhere else, they will abandon you!

Only about one-third of people could be considered "Price Shoppers." The other two-thirds will buy what they want as long as the price is fair or a good reason for a higher price. Focus on the higher spending customer groups first, and you will automatically attract the "Price Shoppers." They will become curious when they see many people buying from you. You can attract and target them with items on your menu that offer. (Ever wondered about the $0.99 menus?). Here are some themes you can use to center your advertising around. All these will attract the right kind of customers:

- Your name.
- Your quality.
- Your specialty foods.
- Your low price.
- Your unique products and tastes.
- Your speedy service.
- Your first-class ingredients.
- Your satisfaction warrant.
- Your generous portions.

These features will position you in the marketplace as a better option and give people reasons to make you the first choice before your competition. The power of advertising lies in the fact that you can influence the opinion of the public mind! You can give people reasons to come to you! Ultimately, you have the power to tell people what to do!

- Communicate your messages!
- Tell people where you are located!

- Tell them what you have to offer!

- List your best features as you believe in them!

- Give them at least one reason why they have to stop by!

- Don't be afraid to use bold headlines and lettering!

- Remember to keep your message as simple as possible!

- Repeat your ads frequently so they can stick to people's minds!

- Once you have attracted the right customer base, you have the best advertising tool available to you: word-of-mouth advertising!

Here is the best marketing advice of all:

- Every customer that had a good experience with your overall service will talk to several people and recommend you for free!

Chapter 8: Make Your Business Profitable in the Long Run

The Launchpad is ready to release and it is time to rev up the engines and stoves of the food truck. Marketing is essential to keep any business running. You should help the business to get noticed so that you can lure in customers. Competitors in the same field of business are never going to rest and make it easier for you. You must advertise and market yourself and your food product efficiently. Here are some marketing tips for the food truck business:

Set Up Weekly Specials

After the launch, you must gain speed and traffic in business. If a customer likes a specific food item like a Mexican taco, you could have "Taco Tuesdays," where you serve the customers tacos at half the normal price. This will spread the word and assure you of many crowds.

Be One with the Community

Get close with the community you want to serve. Sponsor for a local sports event or try helping with a charity. Also, find ways to tie up with other business owners in the community.

Hold Contests

People love contests and they are an excellent idea to promote your food truck business. Promote the contests through social media and other forms of advertising.

Celebrate Often

You do not need a big reason to celebrate. Opt for smaller holidays and make things exciting and new for the customers. Show the spirit of your celebration through the food you offer.

Have an Inner Circle

Treat your most valuable customers nicely and create an inner circle with them. Offer them discounts and earn their trust by being sweet and nice to them.

After all this, it is also important that you choose the perfect spot to put up the food truck. Make sure that you choose a place where there will be many hungry people. Park your vehicle next to a commercial or industrial space.

Also, make sure that no serious competitors are around to spoil your day. When you want to choose a place, also find out about the events that might happen regularly at that place. Try to participate in such events and maximize your profit in doing so. Be sure to find out how easy it is to get licenses to operate your food truck at these events. Do not feel bad about partnering. Partner up with a mall or building complex that will allow you to set up on their property.

Tips for Sustaining the Successful Run After Setting Up

Feel Free to Market Yourself

It is essential to keep the business running smoothly and controlled. This will make your brand profitable in the long run

Marketing extends beyond the beginning phase, and it is essential to keep the food truck running. Take advantage of digital media and their marketing platforms. Tweet about the places you will put up the stall, connect with Facebook and maintain a Facebook page to post regular updates. Have a well-planned social media marketing scheme and try to lure in more customers by showing the merrier sides in dining with you. Also, make sure that you offer the quality and service you have advertised. False advertising can easily ruin the whole process.

Think Freely and Do Not Attach Yourself to an Idea

Even if you have found the perfect spot for business and even it has worked well for a long time, there is a possibility of dwindling sales. Take time to re-plan and think about moving to another new area. Do not be too rigid in the way you think. It is a waste of time, and you might end up losing the business in the process.

Expand on the Revenue Streams

Change with time and try implementing new business ideas. Take risks and always be on the lookout for new opportunities. Cater to events and festivals to increase the profits you take. Get out of your comfort zone and try new and exciting things. Keep the energy and the flow running.

Be Open to Teaming up

Do not feel bad about teaming up with other food truck owners out there. You could get a lot out of it because people who eat out of food trucks are most likely to change their trucks often. Pick a crowded place and a friendly food truck owner to club your business with. Cater to that crowded place together and get the best out of that situation. It need not be done regularly, but it is good to team up occasionally. People will also love the variety that you and your friend in business offer.

Keep Networking

Make friends with people who have a strong influence over the place. Drop the prejudice and consider asking other truck owners to get valuable referrals for events and festivals. People might help you, and you might even expand your network. Do not live in your world and miss the exposure that others offer to you.

Make a Good Investment in Your Staff

Make sure that you help the staff grow within their positions to stay trustworthy and faithful in the future. You must treat them with the respect they deserve, and you must acknowledge their good work. The process of bringing in and training new staff is time-consuming and costly.

Put a Good Price Tag on Your Food Items

Even if you are new in the business, you necessarily offer food at a very cheap rate. So, if your food is tasty and has the excellent quality, feel free to charge the price that will benefit your system. It is vital to remember that people are ready to pay for good things. Keep your eye on the quality of the food you serve, and you will see your business grow automatically.

These tips and techniques are essential for your path to becoming a successful food truck owner. So, get out there and put some interesting items on the menu to keep hungry taste buds on fire. Serve with a bright smile on your face and complete love in your heart. There are many people to feed in this world, and it is high time that you realize that you can be the change you want to see. Thrive and work hard to serve the tastiest food on wheels and make sure that you touch people's lives with what you do.

Chapter 9: Maximize Profits and Minimize Costs

While a food truck has the advantage of moving to where the customers are, they also have the disadvantage of breaking down. Even the slightest problems require a reliable mechanic to fix problems before they arise.

Almost unlimited mechanics are available for most vehicle repairs, but generally less for food truck repairs, depending on the problem and location. That's why it is important to be proactive and find a few reliable truck mechanics before you run into problems.

When looking for reliable truck repair, consider speaking with friends or even competitors that you can trust. Search for references of those who use certain mechanics regularly to find someone who is honest and reliable. Also, check online feedback when possible.

In the case of kitchen problems, it is best to talk again to trusted competitors or suppliers who know someone who can help you repair the specific equipment you're having issues with.

Making a Profit

Whether a food truck is born from the mind of an aspiring chef or an entrepreneur, making a profit is the key to success and often why they get into this industry. But if you think that the profits are just going to come rolling in as soon as you open, you'd better think again! Successfully owning and maintaining a food truck generally requires long hours, a lot of competition, and a host of legal issues that vary from state to state. You may have seen popular food trucks in your city or watched television programs that portray the glamorous side of the mobile food industry. Seeing these examples might make you think how hard this can be. You don't see the work each of those owners had to go through to get to where they're at now.

Starting a food truck is like starting any business. You have to build it up from nothing. And the first couple of years is probably going to be the hardest you've ever worked just to scrape up an income to pay for your initial expenses.

This is not to scare you away from starting your own food truck business because profits are made. According to Off the Grid's Matthew Cogen, most trucks earn $250,000–$500,000 each year. Industry experts estimate significant growth in this multi-billion-dollar industry.

Surviving the Winter Months

For most gourmet food trucks, owners must face one obstacle food trucks every year, unless you happen to live in a warm-weather region. Food truck owners will often tell you that they find themselves struggling during the winter months. Unlike a brick-and-mortar business, the number of customers you have during the winter months will dwindle to almost nothing. When your food truck business seems to be slowing down to conditions where it doesn't make sense to even open, there are several alternative venues to consider to keep your business afloat during the cold months.

One option would be catering. Instead of searching for customers in the cold, you should look for clients or events to bring customers to you. Start advertising as a catering business during the summer months and then reach out to cater parties and events year-round.

In addition to parties and events, try partnering with offices and other businesses to serve their needs. Look for tournaments and conventions that need catering. The winter months may even be more profitable for some food truck owners than the summer months. For those who are still willing to work outside, consider mixing up the menu during the winter months. Bring in coffees and hot chocolates to the menu to boost sales, keeping your customers warm during the off-season. You should embrace the mobility of a food truck. There is no reason to stay in one location. Move to businesses that need catering or extra choices during the winter months and look for big events near you that could use additional options in unique and quality meals.

In the end, it is your decision whether to stay open during winter. Many other businesses make their money in the summer months and take winters off. But if you do close down for the winter, you need to keep focusing on improving business and profits the following season.

Unexpected Expenses

Just like the basic expenses of a restaurant, food trucks have other expenses to consider. With gas prices consistently increasing, your fuel costs can add up to be a hefty expense.

When parking your truck for business, finding a profitable spot can also often result in receiving a parking fine if you're not careful. Parking in the wrong location can result in massive expenses, with parking tickets each costing up to $130 or more. Some food truck owners consider this the "cost of doing business," but generally, it is not a good practice. Weather conditions can also cause problems for food trucks. If customers are hoping to stay out of the rain, they may also avoid your truck, despite passing by your truck on their daily routine.

In addition, storage space can increase more and more throughout the year. Even vendors who cook on-site will need to prepare certain ingredients in a commissary or commercial kitchen, which often requires additional storage facilities.

Chapter 10: What's Your Best Food Truck?

Now it's time to talk about your truck because, after all, it is an integral part of your business! Your truck differentiates you from brick-and-mortar restaurants and other mobile food businesses. But finding a truck to fit your needs can be like finding the right piece of real estate! The right truck for you will be determined by how many customers you plan to feed and the type of food that will be served. Your budget will play a big part in the decision as well as the location you'll be operating in.

You need to examine two major areas when purchasing a food truck. You will have to look at the vehicle itself, and you have to evaluate the kitchen area. Together, these factors can create a wide range of costs. A food truck can cost anywhere from $15,000 to over $100,000! But why is there such a price difference? Here are some factors that influence that number:

- Buying new vs. used.

- Vehicle size.

- Onboard equipment.

- Retrofitting.

If you're buying a used truck that needs retrofitting, you can realistically expect up to a year to complete! As a rule of thumb, it's always a good idea to estimate a longer completion time to avoid disappointment. After all, how many times have you heard a contractor say it'll take X number of days, but in reality, it took much longer than their promised completion time? On the other hand, a brand-new food truck can pretty much be ready to go into operation right off the lot. Startup time is going to be a lot quicker with a new truck!

Advantages of New Trucks

There are several advantages to buying a new food truck. One advantage is that they can be designed exactly for your specifications. Without retrofitting, a new food truck will most likely meet all current vehicle safety and health regulations. Buying new also means you'll have a new vehicle warranty, much like you'd get when buying a brand-new car. This can be reassuring if anything breaks. New trucks cost more than used trucks, but you get the advantage of starting up with all your equipment in prime condition! However, many mobile business owners choose to purchase used trucks mainly because of their lower costs.

Saving Money with Used Food Trucks

If you go the route of buying used, you should have a trusted mechanic inspect the truck thoroughly, just like you'd do when buying a used car. Most trucks have been used in harsh conditions, and you need to be sure of what you're getting into before you lay down any money! When shopping for a used truck, there are some key issues you need to look out for.

The first one is engine reliability. This will directly affect your ability to get to your locations. Another issue is the amount of interior space. You need to determine if it has enough room for your staff and the kinds of foods you want to cook, like buying a used car; you need to look at the make, model, year, and mileage.

If you're going to retrofit, you need to add the cost of retrofitting to the truck's price to figure out the total cost. If the truck comes with existing equipment, do a proper inspection to see if it's in good working order. You may or may not have to replace it. As an alternative, you could end up converting a non-food-related vehicle into a food truck. When using a non-food-related truck (such as a large delivery vehicle), you're going to have to do a total retrofit to turn this vehicle into a food truck. This is probably one of the most time-consuming methods of all!

Whether you're buying new or used, there will be some common concerns that will arise for both. One of the issues to ponder will be whether there is enough room for all your cooking equipment. Will the cost of the truck fit within your budget? Budget constraints always seem to be a recurring theme with any business.

Are You Qualified to Drive a Food Truck?

Another thing to consider is whether you can actually drive the truck safely. Have you ever driven a truck like this before? This will be the first time driving a vehicle in this weight classification for most food truck owners. You will need to know the gross weight because trucks over 26,000 pounds require a commercial driver's license to operate them. Wouldn't you know it? Obtaining a commercial driver's license (or CDL) is yet another item on your to-do list!

Therefore, when shopping for a food truck, you need to plan and review all available options carefully. Remember that your truck is the center of your business and is the most visible aspect of it! Along the way, you'll probably work with a food truck designer who can create a virtual floor plan for your truck. This will help you better visualize the interior space and arrangement of equipment. Like any construction job, your space requirements need to be accurately mapped out before actual work begins! Poor planning can and will result in costly mistakes!

Renting a Food Truck

Now that we've covered new and used trucks, I want to introduce you to one more cost-effective option for prospective food truck owners. This option involves renting or leasing a food truck, which can save you a lot of money, especially when you're just starting! Renting a food truck is great for those who want to try their hand at mobile food but aren't sure they are ready to make this a permanent career. Realistic rental rates can be found at around $2000 a month.

Some people plan to rent a truck when first starting out. Then, when income goals are consistently met, they are ready to buy their truck. Like any piece of machinery, you'll need to perform regular maintenance to keep it running because you can't make money if you can't get your truck to your customers.

Vehicle Storage and Parking

When it comes to storing and parking your truck, there are some requirements you'll need to follow. Most cities require that food trucks park at a Department of Health-approved location when preparing food before or interim between services. However, what types of facilities are approved? Here are some of typical food truck storage facilities:

- Commissary.

- Commercial kitchen.

- Dedicated mobile food facilities.

You'll also need to check your local laws and regulations for overnight food truck storage. This is most likely a commissary. According to Mobile-Cuisine.com, the commissary is the lot that you're legally required to park your vehicle when not in use. It's against the law to store or prepare food, beverages, or any food-related items at a private home or unapproved structures. Commissary costs are paid monthly and could cost on average about $800–$1200 per month. The rate varies because each commissary offers different types of features. They can include security cameras, 24-hour a day security staff, electricity, fuel, or other necessary supplies.

Chapter 11: Food Hygiene and Safety

What is food poisoning? It is an acute illness, generally sudden, realized by eating contaminated or unhealthy food. Food poisoning symptoms vary with the source of contamination. Most types of food poisoning cause one or more of the following signs and symptoms:

- **Nausea:** A queasy feeling as though you were going to be sick.

- Agitation.

- **Agonies in the bowl:** Holding torments in the area of the stomach.

- Diarrhea.

- Fever.

The essential driver of food contamination is:

- **Bacteria:** The commonest.

- **Viruses:** They are more modest than microscopic organisms.

- **Chemicals:** Insecticides and weed-executioners.

- **Metals:** Lead pipes and copper dish.

- **Harmful plants:** Toadstools and red kidney beans (deficiently cooked).

Microbes are the most widely recognized sort of food contamination; along these lines, we must get some answers concerning them. Microbes are unassuming bugs that live recognizable for what it's worth, in water, in soil, on and in individuals, in and on food. A few microbes cause disease. They are called pathogenic microbes. A few microorganisms cause food to rot. They are called spoilage microorganisms.

Warmth

They love an internal heat level of 73ºC yet can happily develop at 15 ºC. They become most promptly somewhere between 5–63 ºC. This is known as the danger zone

Time

Each bacterium creates by separating fifty-fifty, considering everything, at standard spans.

Dampness

They need water, and most sustenance have enough water or sogginess to permit the microorganisms to prosper. A few microorganisms can shape a hard-protective case around themselves, known as a spore. This happens when the 'going gets remarkable' when it gets nonsensical hot or pointlessly dry. This way, they can suffer incredibly hot or cold temperatures and can even be accessible to dried sustenance. At the point when the right conditions (5–63 ºC) return, the spore rises out of its cautious bundling and transforms into a creating, food pollution tiny life forms once more.

Microscopic Life Forms and Food Poisoning

We have set up that microorganisms' presence is one of the most broadly perceived explanations behind food pollution—the presence of destructive manufactured mixes can correspondingly cause food defilement. There are a couple of conceivably unsafe engineered substances present in food. For example, potatoes that have turned green contain Solanine's hazardous substance, which is essentially unstable when eaten in excess.

Rhubarb contains Oxalic Acid—the wholes present in the stems which are reliably cooked are respectably harmless to people. Yet, the higher concentration in the leaves makes them outstandingly dangerous to eat. A poison is a destructive substance that might be conveyed by processing a plant or animal, especially certain tiny living beings. Dangerous food pollution is essentially achieved by Staphylococci in the UK and sometimes in this country, Clostridium Botulinum. Nourishments most ordinarily impacted by Staphylococci are:

- Meat pies.
- Synthetic cream.

- Sliced meats.
- Ice-cream.

- Pies with sauce.

50–60% of people pass on Staphylococci in their noses and throats and are accessible in nasal releases following an infection. Staphylococci are likewise present in skin wounds and infections and discover their way into sustenance using the hands of a spoiled food regulator—consequently, the centrality of keeping all injuries and skin conditions covered. Notwithstanding how staphylococci are quickly crushed through careful cooking or warming, the toxic substance they produce is consistently impressively more warmth safe and may require a higher temperature or longer cooking time for its pummeling.

The substances most affected by Clostridium botulinum are:

- Inadequately arranged canned meat.
- Vegetables.
- Fish.

During the business canning measure, each care is taken to ensure that each piece of the food is warmed to an adequately high temperature to ensure pulverization of any Clostridium botulinum spores that might be accessible.

Yeasts and Molds

Little living things, some are alluring in food and add to its characteristics. For example, the development of cheddar, bread maturing, etc. They are essential plants that seem like hairs on food. To create, they require warmth, sogginess, and air. They are executed by heat and light. Molds can develop with too little suddenness for yeasts and minute life forms to create. Yeasts are single-celled plants or creatures bigger than bacteria that produce their sustenance from sogginess and sugar. Sustenance having somewhat sugar levels and a broad extent of fluid, for instance, regular thing presses and syrups, are liable to age considering yeasts. Yeasts are crushed by heat.

Contamination

Tiny particles imparted by food, which may cause infection. For example, Hepatitis A (jaundice). As opposed to microorganisms, contaminations can't add or fill in food.

Protozoa

Single-celled day-by-day schedule structures with water involvement and are subject to certifiable disorders, for instance, wild fever, generally spread by polluted mosquitoes and looseness of the bowels. These food-borne sicknesses are, by and large, gotten abroad.

Escherichia Coli

E. Coli is a regular piece of the assimilation plots of man and animals. It is found in human excreta and rough meat.

Salmonella

It is accessible in the stomach-related organs of animals and individuals. Sustenance impacted fuse poultry, meat, eggs, and shellfish.

Control of Bacteria

There are three strategies for controlling minute living beings:

1. Shield food from microorganisms detectable all around by keeping sustenance covered. Use

separate sheets and cutting edges for cooked and uncooked sustenance to prevent cross-pollution. Use different concealed leaves for explicit maintenances. For example, red for meat, blue for fish, yellow for poultry, etc. Store cooked and uncooked sustenance freely. Wash your hands frequently.

2. Take the necessary steps not to keep sustenance in the danger zone of someplace in the scope of 5–63 ºC for more than should be normal.

3. To wipe out microorganisms, subject organisms to a temperature of 77 ºC for 30 seconds or a higher temperature for less time. Certain microorganisms are structured into spores and can withstand higher temperatures for more widened periods. Certain artificial materials, in like way, kill tiny creatures and can be used for cleaning stuff and utensils.

The basic food neatness rules of criticalness to the cook are Food Safety (General Food Hygiene) Regulations 1995 and Food Safety (Temperature Control) Regulations 1995. These executed the EC Food Hygiene request (93/43 EEC). They replaced a couple of interesting rules, including the Food Safety (General) Regulations of 1970. The 1995 Regulations are indistinguishable in various respects to earlier laws. Regardless of the Health and Safety authorization, these rules highlight owners and heads to perceive the security dangers and design and execute reasonable structures to hinder pollution.

Hazard Analysis Critical Control Points (HACCP) or possibly Assured Safe Catering cover these structures and frameworks. The rules place two general requirements on owners of food associations to ensure that all food-managing exercises are done neatly and shown by the 'Rules of Hygiene.'

What's more, there is a responsibility on any food controller who may be experiencing or conveying an infection that could be communicated through food to report this to the business, which may be obliged to forestall the individual worried about managing food. Cooking foundations have a general obligation to oversee, teach, and prepare sanitation and cleanliness comparable with their representatives' duties. Insights concerning how much exercise is required are not determined in the guidelines. Notwithstanding, HMSO Industry Guide to Catering gives direction on preparing, which can be considered a prevailing norm to follow enactment.

Prevention of Food Poisoning

Practically all food poisoning can be prevented by:

- Complying with the rules of hygiene.

- Ensuring high standards of cleanliness to premises and equipment.

- Preventing accidents.

- High standards of personal hygiene.

- Physical fitness.

- Keeping up great working conditions.

- Keeping up equipment in decent shape and clean condition.

- Utilizing separate equipment and knives for cooked and uncooked foods.

- Ample arrangement of cleaning facilities and equipment.

- Put away foods at the correct temperature.

- Safe reheating of foods.

- Quick cooling of foods before storage.

- Protection of foods from vermin and insects.

- Hygienic cleaning up procedures.

This has been only a brief overview of food safety. If you are in the catering trade or are arranging to become a cook or chef, you must learn everything about the subject. The simultaneous connections should help to fill the holes.

Chapter 12: Organization of Space and Interior Design

It's also important to make your truck stand out from the rest. You can do this through aesthetics.

While some people say that food is the only important thing in any food business, it isn't true. Of course, it's also important for customers to be able to eat somewhere nice because no one wants to eat in a truck that's rusty or that's not even designed at all. If you don't have time to set up your truck so that it would attract people, it may also mean that you are not yet ready for this business and that you may have to think things through. Anyway, some things you have to keep in mind when it comes to designing and decorating your food truck. These things are:

The Theme

Suppose you're creating a burger business. It won't be right to use pastels as the theme or put photos of classic Hollywood stars on the walls of your truck, would it? You have to make sure that the theme you choose is connected to what you're serving so that your customers won't be confused.

Color Scheme

The main rule is to use the colors on the opposing sides of the color wheel. This way, everything will go together, and your truck won't look like it's painted by a two-year-old. Also, it would be nice if the color scheme of your truck is also something you can use for the uniforms of you and your staff to make everything cohesive.

Seats

Some food trucks allow their customers to sit around the truck, so if you can put out some chairs or anything that your customers can sit on would be good.

Utensils and Packaging

It would also be nice if you could set up the truck so that your customers won't have a hard time getting the utensils they need. Always keep condiments and tissues around because most customers need them, and make sure that you have environment-friendly bags that they can just pick up and put their orders in so they can take them on the go.

And of course, give it some life. The best thing you can do with your truck is put some of your personality in it. This way, your truck won't be generic, and when people see it, they'll be excited to eat. When people notice that a food truck has life and that it's something cool, chances are they'll go on and try your products—and that's good for you! Attract customers, and they certainly will eat what you have prepared! Let your truck speak for itself.

Chapter 13: What Machinery to Buy

The first selection is to choose the food truck in which you will do business. You should select the best possible machine to meet all your food-dispensing needs. Before selecting the vehicle, you must check the local codes and restrictions on the vehicle's sizes and where they ought to be parked.

Assure that you have assessed the business needs and be confident about the choice. If you are going to make food on the spot, you will need a big truck. Ensure that you study the business needs completely to determine the amount of space required to run the business. Sometimes you have to serve cold sandwiches and prepare pizza and fries on the spot. Assure that you have enough space for the same.

Set your budget and ensure that the financing is secure. Stick to a budget and get all the vital things in keeping the business running. Do not go overboard in using the money, and do not make stupid investments in unnecessary equipment. Make sure that your financing is secure and safe before even thinking about buying the truck. Take time to do it, as bank loans can take a long time to get processed.

Seek Professional Help Before Driving to Different Places

Have a practicing mechanic with you when you buy a truck. Ensure that the mechanical components are intact and in the best working conditions. The truck and the machines are the blood and bones in this business. They must be reliable. Do a complete checkup of the engine, transmission, and drive train. You can also talk with other expert food truck owners for their advice, and it might help you in a very prudent way.

In selecting the staff to run the business, it is advisable to find potential catering employees. They are the right kind of people to run this business. Place an ad or put a word out so that candidates find interest in working with you. It is best to hire outsiders and not your close friends and family, as it can lead to conflicts later. Set up interviews after the shortlisting of candidates. Set a day aside and intimate them in advance about the interview. Make sure that you read their resumes in advance and make sure that they have decent communication skills. It is also advisable to look at their physical appearance as an unpleasant worker can repel customers from buying. Do not hire people on the spot. Take time to review all the desired candidates and pick the best out of them. Try to contact their prior employers and ensure that they are good at doing their jobs.

Another critical asset to select and categorize is the equipment for the food truck. If you want your venture to be successful and running, the equipment should be best in class and quality.

Select the Right Kind of Equipment

Different dishes require different equipment. You should go into the details of making the dishes that you want to cater to people. List the tools and supplies you will need to cater to the needs and put every tool down on the list. Add even small things like plastic forks, tin foils, and spatulas. Do not spend on unnecessary equipment's and if you are looking forward to expanding the menu, make sure that you are flexible enough to grow your budget.

Apart from all these, standard equipment is important to have in a food truck. They include:

- Plumbing system.

- Sink for dishes and hand sink.

- Drainboards, grease trap, and disposal system.

- Water heater.

- Freshwater and greywater tank.

- Exhaust hood and fan motor.

- Interior lighting and electrical outlets.

- Fire suppression system and sprinklers.

- Awning or ordering window with glass or screen.

- Covering for walls and ceilings.

- Air conditioner and heater.

- Generator.

- Storage space for supplies, tools, and utensils.

- Storage space for food and ingredients.

- Freezer space.

- Stoves, grills, and fryers.

- Warmer/cooler.

- Food-safe preparation space.

With a stocked-up truck and pumped-up heart, are you ready to serve some tasty flavors to all those hungry tummies? Not yet. Like any other normal business, the food truck business also needs sound marketing strategies to thrive and succeed. Even with the yummiest menu and awesome staff, you will have to know the secrets to keep the business running.

Chapter 14: What Kitchen to Use

Did you know that there are strict laws and regulations regarding overnight storage of your food truck? The most common facility for food truck owners to park their trucks is a commissary. The commissary is where you are required to park your vehicle when not in use.

Commissaries can also be commercial kitchens. The reason commissaries and commercial kitchens are required is because they help keep this industry in check and promote food safety. As a food truck owner, you will spend a lot of time working inside your truck. However, you're more likely to spend an even greater amount of time working in a commissary or commercial kitchen. The primary reason you need the services of a commissary is that it is illegal to prepare the food you will be selling from home. Your food truck and commissary will need to be officially approved and operate within the guidelines of local health codes.

Health inspectors will also check that you use an approved commissary or commercial kitchen. Their main job is to make sure that your food is stored and handled safely, whether it's inside your truck or at a commissary. Any violations can cost you time and money. If you are already a restaurant owner, then you already have a commercial kitchen at your disposal. A commercial kitchen or commissary will need to be added to your list of expenses for those who don't.

Costs of Commercial Kitchens

Commercial kitchens make their money by charging a monthly fee for you to use them, but there are creative ways to help you reduce your costs when it comes to renting a commissary. Commissaries usually charge a monthly fee with average costs that could run you $800–$1,200 per month. This rate varies highly because each commissary offers different types of services. If cost is an issue, a simple way to reduce expenses is to partner with and share a commercial kitchen space with another business. This way, you can split the cost of the monthly fees.

Some facilities can be bare-bones, while others may include security cameras, round-the-clock security staff, electricity, fuel, or other necessary supplies. One of the requirements needed before renting a kitchen is liability insurance; however, each commercial kitchen or commissary will undoubtedly have additional requirements that are unique to each location. Just be sure to get all the specifics before signing a contract; if you have trouble finding an affordable commissary, some other options you should consider.

Commissary Alternatives

Local schools, churches, or other venues may have a certified commercial kitchen that you can rent. Additional options can include hospitals, firehouses, and even catering facilities that already house the types of equipment that you will need. If you decide to go with one of these alternatives, you will most likely coordinate the use times with the kitchen owners or other users. In addition, it's important to understand that you may only have access to these facilities early in the morning or late at night. Depending on your menu and operation times, this might not be the best option for you. But with a little creativity, you can potentially uncover a great deal on a kitchen rental.

If you've located a commercial kitchen you want to rent, make sure everything is legal and the contract is in writing. That way, both parties know what to expect in case issues arise. Having your lawyer look over the contract is also advisable. The various rules surrounding commercial kitchens exist to protect the consumer and promote safety for the food truck industry. You should never place your food truck business at risk by taking shortcuts for cleanliness.

Services Offered by Commissaries

A commissary or commercial kitchen is essential for a food truck business to operate fully. They offer food truck owners services to follow regulations and laws. Not only do you prepare food at a commissary, but also there is a host of other daily activities that are performed there.

Convenient Access to Supplies

A well-equipped commissary will have the essential supplies on-hand that you can get access to in a pinch. So, rather than rush to a store to purchase items you need, you can easily get them at the commissary. The time savings can be huge when things are busy, and when are they not? However, you'll have to check on the fine print on your contract because there may be certain services you are required to pay for when you are based out of a particular facility.

On-Site Storage

One of the big benefits of a commissary is that you can get on-site storage of your ingredients and supplies. Commissaries are fully licensed and approved for commercial food storage. Depending on the amount of storage that you have agreed upon at the facility, you can stock up and save money by buying some of your supplies in bulk without having to worry about where you're going to store it all.

On-Site Parking

Another benefit to a commissary is parking your food truck at the facility. It's not just a convenient place to park, but it is also where you are legally required to store your truck at night. Remember, you cannot park your food truck at your home.

Charging Stations

During your service hours, your truck will most likely run off generator power. But when you park overnight, you will want to connect your vehicle to shore power to keep batteries charged or refrigeration equipment running. Your commissary should have adequate connections to keep your truck plugged in. The cost of the power could be included in your rent or may be charged separately for actual power usage.

Cleaning and Waste Disposal

Another important feature of a commissary is the cleaning and waste disposal facilities. After a food service, typically, a truck will return to a commissary where cooking tools and equipment can be cleaned. There are strict rules regarding how often certain surfaces and equipment need to be cleaned. Your commissary will help you stay within regulations and avoid any violations. Wastewater requires special disposal facilities because it contains grease and food particles that cannot be dumped into regular drains. Any used water and solid waste need to be disposed of at the facility. Heavy fines can be imposed if disposal regulations are not followed.

Vehicle Maintenance

Time is money, and it's never a good time when your truck needs maintenance work. However, many commissaries offer mechanics services on-site. If your commissary provides mechanical services, routine maintenance becomes a whole lot easier. Other services can include kitchen equipment repair, periodic inspections, and other necessary work.

Your commissary is your partner as you work in the food truck industry. It is also a place where you can interact with other food truck owners who are working in close quarters with you daily. Not only are you part of the larger food truck community in your area, but you become part of a smaller, more intimate group at the commissary. Sharing ideas and getting help is a lot easier when you've got the support group to back you up.

Chapter 15: How to Create the Menu of Your Food Truck

The Most Asked Question About Our Menu

More than 50% of new customers ask the same question, "what's the best item on your menu?" Our response is, "everything on the menu is good," then we take the time to explain the menu items and why they are so good. Why do we do this? Well, it is a two-fold process:

1. We believe and know without a doubt that the item on our menu is good because it wouldn't be on there if it weren't.

2. It costs eight times more to gain a new customer than getting repeat business from a customer that already frequents your business.

If we're winning the first-time customer, they will become a repeat customer, thereby growing our business even further. If your product is good when the customer first ate at your food truck and they had a good experience, they will be back. Your job in servicing the repeat customer is to be consistent with the product, meaning it must taste the same as when they first ate it.

Menu Design and Layout

One of the biggest faults I see food trucks and small independent restaurants do is improper advertising their menu items. I have learned over the years of doing this food truck stuff that people order food based on what they see with their eyes. So why would you put many words on your menu and no pictures of what you are selling?

The fast-casual restaurant chain Chipotle Mexican Grill© (CMG) does a superb job of presenting their menu items for ease of customer selection and offers great visibility of orders being processed for customers' observation. Despite some of the food safety incidences CMG has suffered in recent years, people still frequent the restaurant because they see their food being made in front of them. In CMG's defense, they are aggressively addressing food safety with their employees to remedy these issues and win back customers. These people are the case study to follow on how to recover from highly publicized food safety issues. It is better not to have an issue in the first place—plan to succeed! Here are some things to consider as you develop your menu:

- Keep your menu simple with five entrée items or less.

- Remember, time is money; more menu items will slow your quick processing of customer orders. Americans hate standing in long lines; your five menu items need to knock it out of the park in taste and presentation.

- Take pictures of the menu item you are serving to customers in the packaging you are presenting.

- Do not waste money getting professional photos of your food. A patiently photographed smartphone picture is good enough.

- Keep your menu clean and do not number your menu items, you want people to commit their favorite food item to memory (Guess what the number for a Big Mac® at McDonald's® is? Who cares? Everyone knows what a Big Mac is and looks like. Get my point?).

Don't overthink it! You are a food truck, not a restaurant; people come to your business for their convenience. Simplify their process to make ordering food easy. Next, design options:

- Use menu templates from companies like Vista Print. Vista Print's staff is available 24/7 to offer customers advice, and their finished products are professional, neat, and make a statement (I use them for 99% of my advertising and marketing).

- Select a high gloss poster-size menu board (24" x 36").

- Ensure you set aside a segment on your menu board to arrange the pictures of the food you are marketing to customers.

- Glue the posted menu board ordered from Vista Print on a precut poster board purchased from BLICKs, Michaels, or Hobby Lobby.

- Finally, get creative on how you display your newly designed menu board. You can affix it to the side of your truck or use a sidewalk cafe board to hold your menu board.

- Now let's add up the cost; smartphone photos of menu items (photographed by you), Vista Print account (set up online by you), selecting a menu template from the Vista Print Website, and upload SMART Phone photos on your own Vista Print account. The self-organized layout of your menu on the Vista Print Website, confirm proof of your design layout, payment for your Vista Print order online via credit card = total cost $50 for two menu boards.

Money saved, priceless. Technology today is a tool that enables us to have a competitive advantage over impatient people. Invest the time and effort to make a good product. It is your business, and it should reflect you. Do not put out sloppy work. I have faith in you to be great!

How Much Should I Charge?

Menu prices need to be set, at a minimum, 3-times the actual food cost of the food components used to transform the menu item into a final product sold to the customer. Why? Because you need to generate income to cover your business expenses (i.e., labor, fuel, licensing fees) so you can make a profit on the items you sell.

For example, if you sell a ¼ lb. Gourmet Hamburger with ground beef from grass-fed cattle free of antibiotic injections, you need to sell the burgers at a price that will yield a profit for your business. Let's assume the food components of the burger consist of the following: ¼ lb. beef patty, 1 hamburger bun, some lettuce, 1 slice of tomato, 2-GMO free butter chip pickles, a splash of your special condiment sauce, and 1 Styrofoam container to keep the burger in. See the breakdown below on figuring your unit cost and recommended amount to charge the customer.

Total unit cost = $2.45

Recommended minimum menu price = $7.35

The next question to ask is, will your customer pay $8.00 for a gourmet burger when they can buy a less expensive substitute for a third ($2.65) of your offering price? It depends on which part of the country you live in and a market for customers who want grass-fed beef free from harmful antibiotic injections.

Food Component	Calculation	Unit Cost
1-lb Ground Beef at $6.90 per lb.	$6.90 ÷ 4	$1.73
Pack of 12 Hamburger Buns at a cost of $2.65 per package	$2.65 ÷ 12	$0.22
Head of lettuce at a price of $0.39 each portioned for use on 15 burgers	$0.39 ÷ 15	$0.03
4-pack of Tomatoes at a cost of $4 per package, portioned for use on 24 burgers	$4 ÷ 24	$0.17
64oz Jar of pickles at cost of $4.99, portioned for use on 50 burgers	$4.99 ÷ 50	$0.10
Special authentic condiment sauce made fresh at a cost of $6, portioned for use on 40 burgers	$6 ÷ 40	$0.15
Foam sandwich container, at a price of $16 each (300 ct)	$16 ÷ 300	$0.05

In developing your menu pricing, aggressively work on getting your food component cost down by buying in bulk and keep spoilage down by ordering on an 8–10-day food-storage window. Applying these measures to control your food component costs can reduce your unit cost by as much as 4% and will enable you to periodically advertise reduced price specials without cutting into your recommended menu price.

Conclusion

Food Truck is becoming a popular business these days, with many people turning to food trucks to provide their meals. With food trucks providing yummy dishes at great prices, you miss out if you don't have your food truck.

A Food Truck or Mobile Food Unit (MFU) is a mobile restaurant that serves ready-to-eat food. It's usually used for lunch and snacks. They are also called mobile restaurants, lunch wagons, catering trucks, motor kitchens, or just trucks. This type of business has become very popular because of its appeal to customers' eyes and stomachs. You don't need to pay rent for a traditional brick-and-mortar restaurant as long as you have a truck or trailer. You can cook food on the spot. You don't need to employ as many people as you would in a traditional restaurant because of their mobility.

As we end this book, here are some tips to ensure the success of your own food truck business:

- **Choose the right truck:** Make sure you have a truck that will fit your business needs. Do not buy a truck just because it's big. The idea is not to have a big truck, but to maintain it well and in good condition. If you have a small budget, it's better to buy something used before because its cost is usually lesser.

- **Budget:** First, determine your monthly expenses for your business and make sure you can meet those expenses at least half of the time. If you don't, sell some of the items you need to operate your business or increase your overhead to meet those expenses or at least 60%.

- **Don't forget about insurance:** You might think you don't need insurance for a food truck business, but think again. As it is a business, you need to take necessary precautions. You can get your insurance from an online or local insuring company.

- **Mobile phones:** You must have a mobile phone in your truck at all times because you can be contacted by relatives and friends if there are any problems with the business.

- **Charge the truck:** You should not allow anybody to charge your truck unless they know how to do it properly. Ensure that your customers know that it is against the law to charge a food truck. Charging the truck only to have someone take a few dishes from your truck is also not allowed.

- **Set up a blog:** In today's age, people are well informed and enjoy reading about your restaurant activity via the internet even when they are at work or going to college, so you can use a blog for updates and have them refer you to their friends via email.

- **Set up a website:** After having a blog, you can set up a website for customers to visit online and keep track of their orders.

- **Work on your marketing plan:** You must have a marketing plan before starting with a food truck business. It would help if you had a brand name, a logo, and graphics. It would be best if you also had your menu and some product ideas, which can be shared with the public via social media.

- **Place ads:** Advertise your business through word of mouth or print media, but you must choose the right type of information that will be useful to potential customers; the wrong kind will not help you in any way whatsoever

I hope that this book will help you start your own food truck business. Thank you for reading this.

Support my work

Thank you for reading this book. I hope you are now more familiar with the food truck business and its prospects.

If you have the pleasure to support and spread my work, then I invite you to leave a positive review on Amazon.

Frame with your smartphone this QR code and review this book.

Thank you

SCAN ME

www.ingramcontent.com/pod-product-compliance
Lightning Source LLC
Chambersburg PA
CBHW050733030426
42336CB00012B/1539